C0-CFI-038

Proteins

by Jane Inglis

Carolrhoda Books, Inc./Minneapolis

All words that appear in **bold** are explained in the glossary on page 30.

Photographs courtesy of: Bruce Coleman, pp. 8, 19, 22; Ecoscene, p. 9; Eye Ubiquitous, p. 11; Chris Fairclough, pp. 14, 17; Science Photo Library, pp. 21, 25 (bottom), 26; Wayland Picture Library, pp. 6, 23; Zefa, pp. 5, 7, 10, 12, 18; Zul, p. 25 (top).

This book is available in two editions.
Library binding by Carolrhoda Books, Inc.
Soft cover by First Avenue Editions
241 First Avenue North
Minneapolis, Minnesota 55401

First published in the U.S. in 1993 by Carolrhoda Books, Inc.

Library of Congress Cataloging-in-Publication Data

Inglis, Jane.
Proteins / by Jane Inglis.
p. cm.
Includes index.
Summary: Describes the different kinds of proteins, where they are found, and their importance to human growth and nutrition. Includes activities and recipes.
ISBN 0-87614-780-5 (lib. bdg.)
ISBN 0-87614-607-8 (pbk.)
1. Proteins in human nutrition—Juvenile literature.
[1. Proteins. 2. Nutrition.] I. Title.
QP551.I48 1993
92-26759—dc20

92-26759
CIP
AC

Printed in Belgium by Casterman S.A.
Bound in the United States of America

1 2 3 4 5 6 98 97 96 95 94 93

Contents

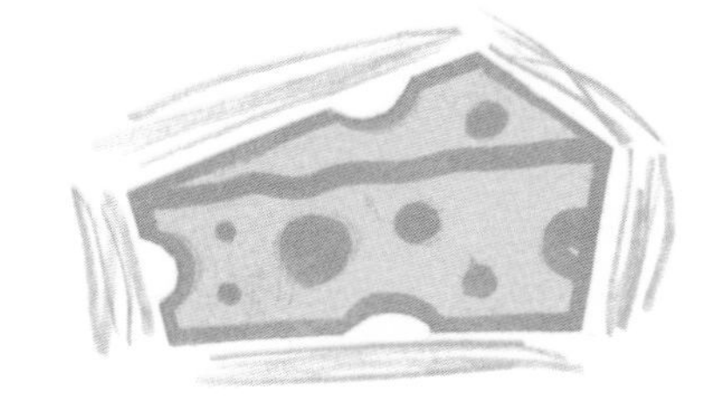

Protein

OPPOSITE
Everyone needs the right amount of protein to grow up healthy. But every person is different, so the right amount of protein is different for each of us.

People need food to stay alive. Children also need food to grow. There are many different useful things in food. They are called **nutrients**, and there are five main kinds of nutrients your body needs. They are protein, **fat**, **carbohydrates**, **vitamins**, and **minerals**. Our bodies need all of these nutrients to do different jobs.

This book is about one of the most important nutrients. Every human being needs the right amount of protein to grow up strong and healthy. This book will tell you how much protein you need and how to get it from your food.

What is protein?

All the nutrients are important, but protein is the most basic of all. Our bodies could not grow without it. Your muscles and all the parts of your body are built with protein. Protein also helps you fight disease, and it helps carry oxygen through your body. Without protein, people die.

Although we usually talk about protein as if there were only one kind, there are actually many kinds of protein. Our bodies need many of these different kinds of protein to build the body and keep it healthy. The protein that carries oxygen is different from the proteins that are made into muscles.

adidas

Every living thing is made up of millions of tiny parts called **cells**. Every cell has proteins in it. Even when you are grown up, your cells are changing and using protein all the time. Children need even more protein than adults to make new cells as their bodies grow.

You can see some of the new cells your body makes if you look at your hair and your nails. These keep growing all the time. Your body uses the protein from your food to make new nails and new hair. If your hair and nails are healthy, you can be sure you are eating enough protein.

About one pound out of every five pounds of your body weight is protein. Proteins are found not only in your hair and nails, but also in your skin, muscles, and bones.

ABOVE *All of these foods contain protein. We need to eat food from many different protein groups in order to get the best mix of proteins for our bodies.*

Some other kinds of protein keep your blood healthy and help your body use the food you eat.

Your body can store some of the nutrients it gets from food, so you don't need to eat more of those nutrients every day. But your body cannot store extra protein. You need to eat a fresh supply every day.

RIGHT *Your body uses some of the protein from your food to make new hair.*

Science Corner

How much of your weight is protein? To find out, weigh yourself on a scale. If you are wearing clothes, take about 2 pounds off your weight. Divide your body weight by five. The answer is roughly the weight of protein in your body.

What Are Proteins Made of?

All the different kinds of protein have been studied for years by experts. We know quite a lot, though not everything, about what proteins are made of. The building blocks that make up proteins are called **amino acids**. At least 22 different amino acids are now known. These can be arranged in many ways to make the different kinds of protein.

How many different kinds of protein can be made from these 22 amino acids? Think of amino acids as being like the letters of the alphabet. All the words in the English language are made up of these letters, arranged in different ways. Or think of 22 different kinds of beads, which could be strung together to make lots of different necklaces. All the

BELOW *These nuts contain some of the proteins we need to keep our bodies healthy. We cannot get all our proteins from nuts. We need other proteins, which are found in other foods as well.*

proteins that do all the different jobs needed by your body are made up of the basic amino acids, arranged in different ways.

Adults' bodies can make most of the amino acids for themselves. But there are eight amino acids that their bodies cannot make. These amino acids must come from food. These eight amino acids are essential—something we must have. They are sometimes called the **essential amino acids**, or EAAs. Children need an additional two amino acids while they are growing.

The story of protein is really the story of how we get these essential amino acids from the food we eat. Experts have studied this process for years without agreeing about all of the details. Studying foods and people is difficult. People are not all exactly alike. Different people need different amounts of protein, and the same person may need different amounts at different times. All over the world, people eat very different kinds of food and get their protein in many different ways.

ABOVE *This strand of beads is like a protein. Each bead stands for an amino acid. Together, they make one type of protein.*

Science Corner

Try this experiment with a big group of children. Twenty-two is the best number!

1. Everyone chooses a shape, such as a square or a star, and a color. No two should have both the same color and the same shape.
2. Everyone colors 10 copies of their shape on paper and cuts them out.
3. Mix everyone's shapes in a pile.
4. Each person takes 10 shapes from the pile.
5. Arrange your shapes in rows. How many different ways can you arrange them?
6. Thread your shapes on a string to make a model of part of a protein.

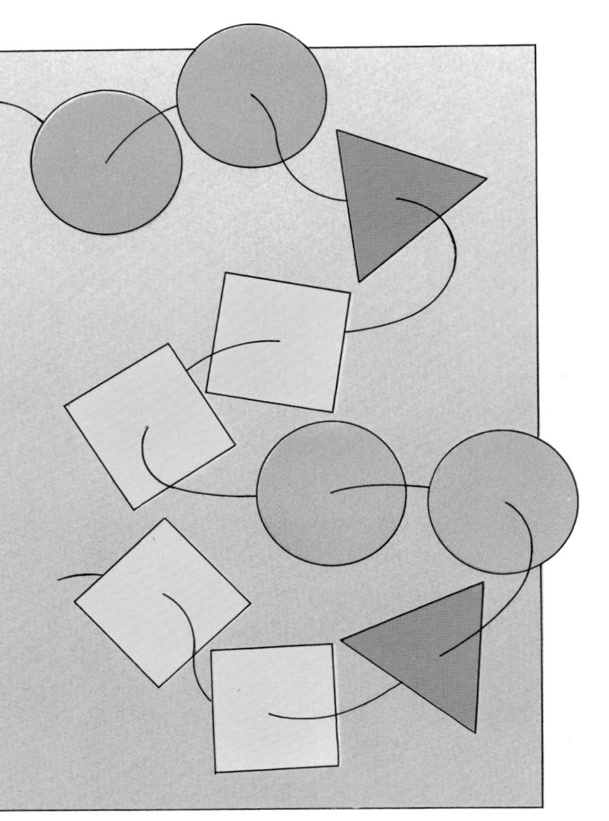

Some people want to change what they eat so that they are healthier. You probably know someone who wants to change what he or she eats in order to lose weight. You might begin to think that people in some parts of the world have too much food or use too much of the world to grow their food. We can all help by eating foods that are better for the planet.

The more we know about food, the easier it is to decide what we want to eat.

A Bunch of EAAs

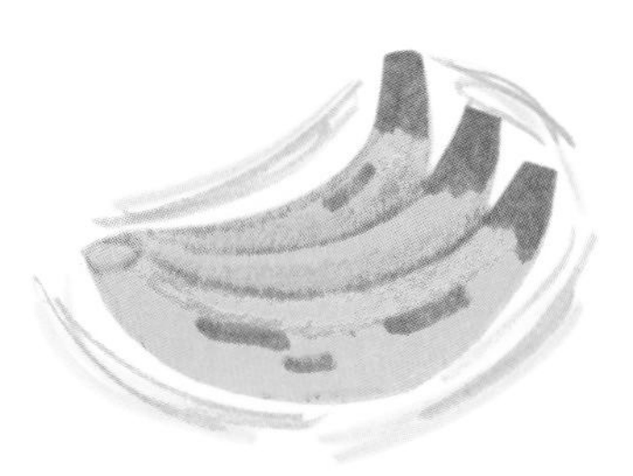

Your growing body needs to get the 10 essential amino acids every day. Not all of the proteins in food give us exactly the right amounts of these EAAs. If there is too little of one EAA, your body cannot use all of the others. You have to get all 10 EAAs in the right amounts in order to be able to use them.

To help you understand this, imagine someone building a house. The house has been planned to use so many bricks in the walls, so many shingles for the roof, so many doors, and so many windows. If the builder had twice the right number of bricks and half the right number of shingles, the house could not be finished. The roof would be half missing and half the bricks would not be used. The doors and windows would be wasted because the house could not be used.

OPPOSITE *In China people get their protein from rice, vegetables, and some fish or meat.*

ABOVE *A builder needs the right materials in the right amounts to build a house. In the same way, we need exactly the right amino acids in the right amounts to build a healthy body.*

Science Corner

To show what happens when you do not get the right mix of essential amino acids, you can do this quick experiment.

Take one of the strings of shapes you made for the last science corner. Imagine it is the pattern for a protein needed by your body.

Now add on any number of extra shapes. The extras will all be wasted because they are not needed for the pattern.

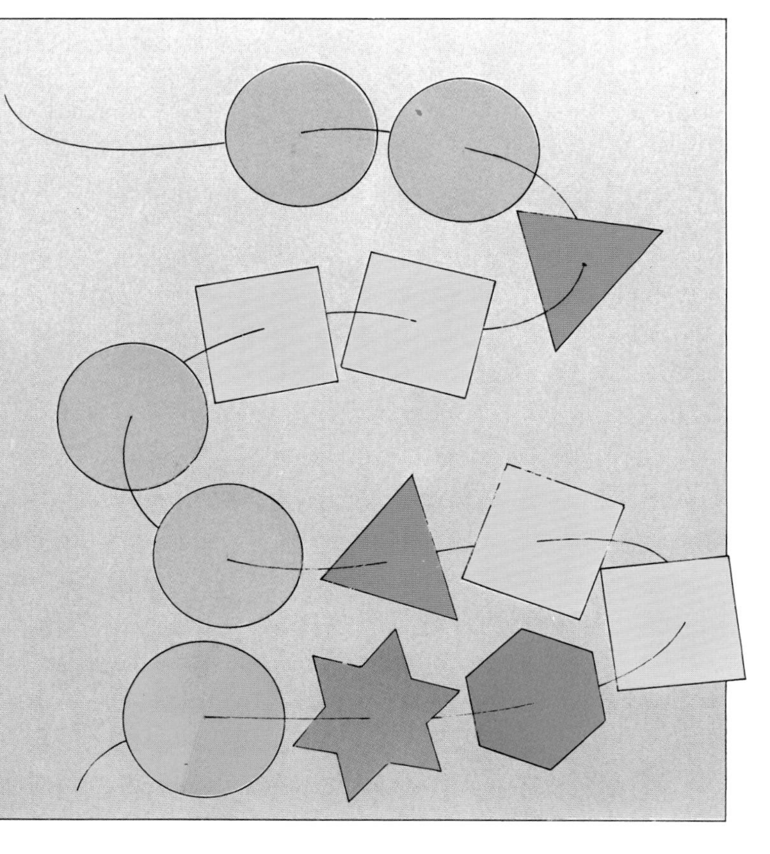

The builder needs the right numbers of doors, windows, bricks, and shingles to finish the house as it is planned. In the same way, when our bodies get protein from the food we eat, they need EAAs in the right amounts.

Of all the foods that give us protein, eggs have the best mixture of EAAs for our needs. All the protein in an egg can be used by your body because there is the right amount of each EAA. This does not mean that you should eat eggs as your only way of getting protein!

BELOW *Eggs contain the best mixture of essential amino acids (EAAs) for our needs.*

It would be very boring to eat so many eggs,

and it would be bad for your health. Like all foods, eggs contain a mixture of nutrients. Protein is only one of the nutrients in eggs. Eggs also contain fat, so much fat that one egg a day is probably too many.

In any case, there is no need to eat only one kind of food just because it contains the right pattern of EAAs. You can eat different foods together so that the mixture contains the right amounts of all the EAAs. This means that you can get all the protein you need by eating lots of different foods. People all over the world have found many ways of eating that use a mixture of foods to get good quality protein.

How much protein do we need?

This question has been answered in many different ways. We do need different amounts of protein at different times in our lives.

BELOW *We need different amounts of protein at different times in our lives. Young children need lots of protein in order to grow.*

ABOVE *These foods contain the five most important nutrients: protein, vitamins, minerals, carbohydrates, and fats.*

Children need lots of protein because they are growing. Teenagers also need plenty. Pregnant women need enough to feed the growing baby, and nursing mothers need lots of protein to help make milk. We all need more when we're sick or when life is pushing us hard. But athletes and other people who lead active lives don't need extra protein. In fact, some experts think that it is bad for us to eat too much protein.

Protein overload?

Once the body has taken all the protein it needs from the food you eat, it can burn up any extra protein as **energy**. But your body also gets energy from the carbohydrates and fats you eat. These foods cost less than proteins, so it is wasteful to eat protein just for energy. It may also be hard on your body. Some experts think your body must work too hard to change a great deal of protein into energy.

Eating large amounts of protein helps you grow quickly early in your life. People who eat lots of protein all their lives may show signs of old age much too soon. Some may live a shorter life than if they had eaten less protein. Cancer and other killer diseases occur more often in people who eat very large amounts of protein. Table 1 shows the average amount of protein needed by people of different ages.

To use this table, you need to understand the idea of an average. If you have a group of six children between seven and ten years old, they will each weigh a different amount. To find their average weight, add up all their weights. Then divide by six. The answer is the average weight for those children.

Tables like this always use average amounts. You may be much bigger or much smaller than most children your age. If you are, you will need a little more or less than the amount of protein shown in the last column. It does not matter if your weight or height is different from that shown on the table. All people are different. This table can help you figure out how much protein you need. It does not say how much you should weigh.

Table 1

Recommended daily amounts of protein

	Average Height (inches)	Average Weight (pounds)	Protein (grams)
Children 4 to 6 years	45	44	30
Children 7 to 10 years	53	62	34
Teenagers (male) 11 to 14 years	61	99	45
Teenagers (female) 11 to 14 years	62	101	46
Teenagers (male) 15 to 18 years	70	145	56
Teenagers (female) 15 to 18 years	65	120	46
Adult males	71	154	56
Adult females	65	120	46

Science Corner

Divide into small groups. Each group then works out the average weight and height for that group.

1. Weigh everyone in the group. Measure the height of each person in the group in inches. Carefully write down your results.
2. Add together the weights of everyone in your group. Divide your answer by the number of people in the group. This is the average weight of the group.
3. Add together the heights of everyone in the group. Divide this answer by the number of people in the group. This is the average height of your group.

Compare your average weights with the numbers in the table on page 15. Decide how much protein each person needs.

Name	Height (in inches)	Weight (in pounds)
1. John	51 in.	61 lb.
2. Sue	52 in.	66 lb.
3. Samara	53 in.	57 lb.
4. Mita	50 in.	63 lb.
total:	206 in	247 lb.

Average height
$4\overline{)206}$ in. = $51\frac{1}{2}$ in.

Average weight =
$4\overline{)247}$ lb. = $61\frac{3}{4}$ lb.

Which Foods Contain Protein?

Nearly everything you eat contains some protein. Plants make protein from water, **carbon dioxide**, and **nitrogen**. The last two things are gases found in the air. All animals get their protein by eating plants or other animals. Table 2 on page 18 shows the amount of protein in some foods.

ABOVE *This squirrel is dining on a nut, which contains protein. All animals get their protein by eating plants or other animals.*

Table 2

Weight of protein in each 100 g of food

One hundred grams is about the same as 3½ ounces or ⅓ of a cup.

	Weight/ 100 grams		Weight/ 100 grams
Apples	0.3	Parsnips	1.7
Bacon	11.0	Peanuts	28.1
Beef	14.8	Peas	5.0
Brown rice	2.5	Potatoes	1.4
Butter	0.5	Soy flour	52.0
Cheese	25.4	Spinach	2.7
Chicken	29.6	Whole wheat bread	9.6
Eggs	11.9	Yogurt	3.6
Milk	3.3		

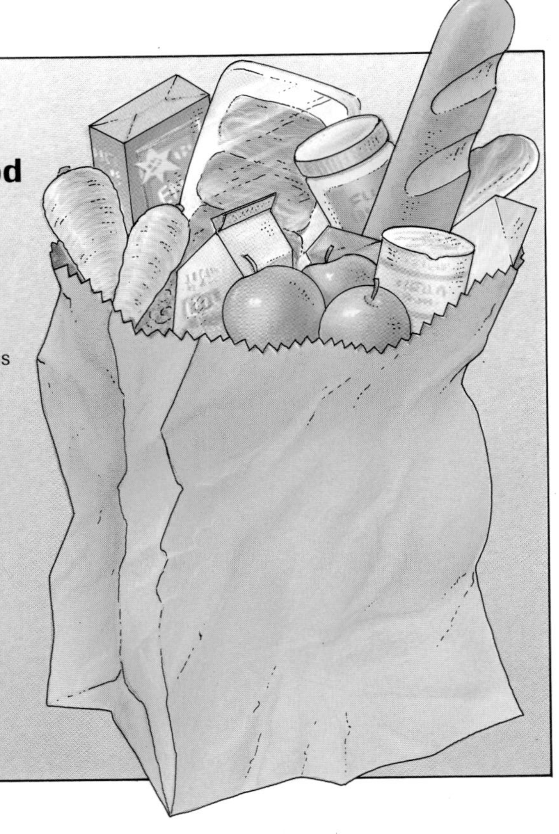

There are some surprises in this table. Some people think you have to eat lots of meat to get enough protein. This is not true. You can see that soy flour actually contains more protein than any of the meats on the list. Most vegetables also contain quite a lot of protein.

Soybeans come from a family of vegetables called legumes. All peas and beans belong to this family. These foods have large amounts of protein and can be grown in places all over the world.

Grains and cereals are also useful for their protein. This table shows that you can get enough protein in your food without eating any meat at all.

BELOW *Some of the fish we buy has been farmed on special fish farms.*

Growing Protein

Some of the foods in the table on page 18 come from plants, and some from animals. Most of the meats on the list contain more protein than most of the plants. This is because animals store lots of protein in their muscles from the foods they eat.

Animals that are raised for meat must eat grain or grasses for months or years to build up those protein-filled muscles. It takes land to grow the food that is fed to the animals. It takes much more land to raise animals for meat than to grow vegetables.

Every day there are more people in the world. The world needs to produce more food each year to keep all the people alive.

People argue about the best ways to use the farmland of the world. If farmers grow the right foods and farm well, we can grow enough food to go around. People will not have to take over all of the wild lands to raise crops. People will not have to damage the earth by trying to farm in deserts or on steep mountains. We can keep some land wild for the animals and plants that have always lived there.

ABOVE *Rice fields produce large amounts of food on a small area of land.*

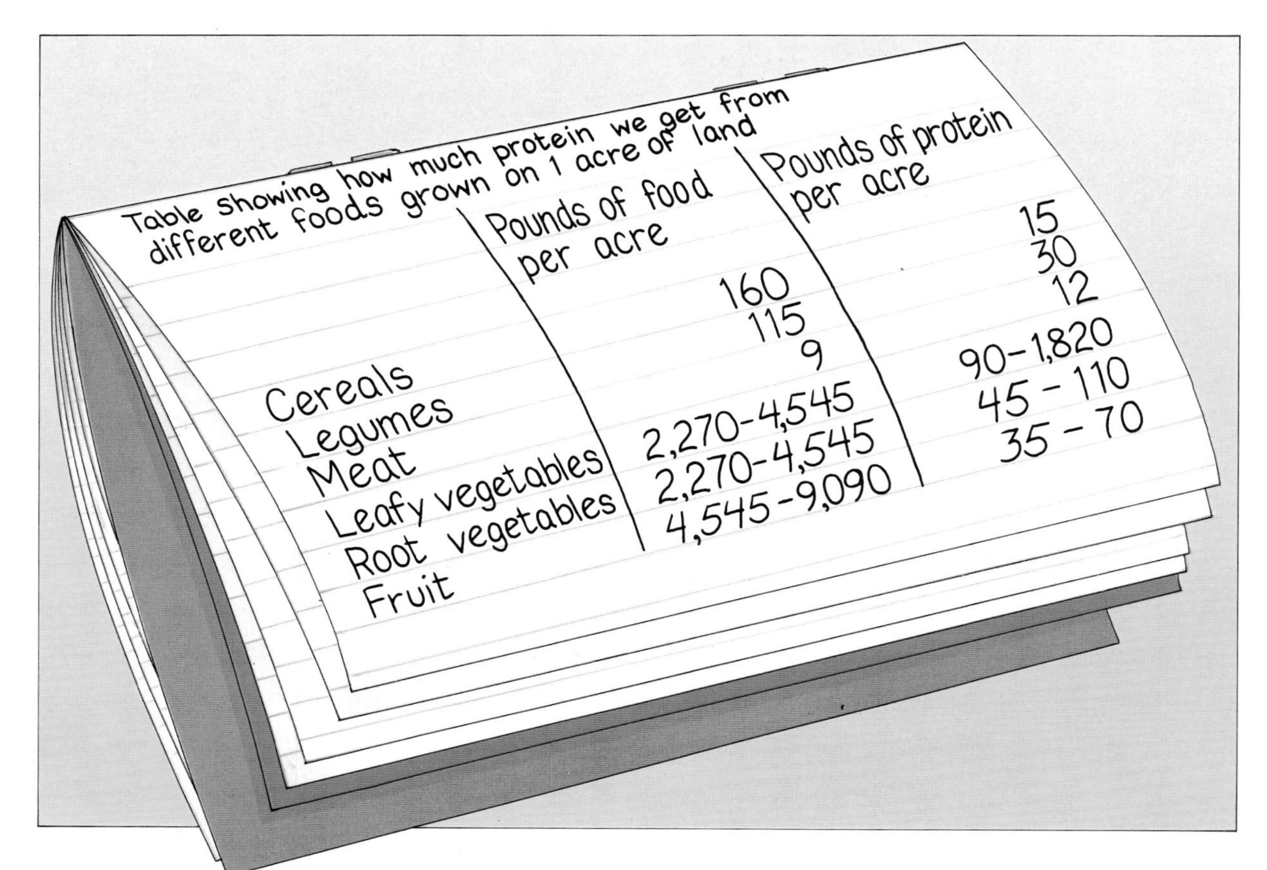

Table showing how much protein we get from different foods grown on 1 acre of land

	Pounds of food per acre	Pounds of protein per acre
Cereals	160	15
Legumes	115	30
Meat	9	12
Leafy vegetables	2,270-4,545	90-1,820
Root vegetables	2,270-4,545	45 - 110
Fruit	4,545-9,090	35 - 70

Table 3 shows different kinds of foods. For each kind, you can see how much food is grown on each acre of land, and how much protein we get from the food grown on each acre.

It can be very wasteful to use good farmland just to raise animals for food. The farmland could give us more protein and other nutrients if it were used to grow plants, too. Some people believe the only way we can feed everyone in the world in the future is by using all farmland to grow plants instead of meat. They believe that everyone will have to give up meat so we can raise enough food.

But there are places in the world where crops cannot or should not be grown. In many places, farming damages the earth. In others, too little rain falls, or the soil is too poor to grow food. Yet people have lived in these areas for thousands of years. They graze animals on the plants that grow wild there, or they hunt the wild animals that live there. Meat is the best way for these people to get protein and nutrients.

People have always eaten meat. It can be part of a healthy diet. But eating less meat may be better for our health and for the rest of the world as well.

BELOW *This meal of chicken and vegetables is well balanced. It contains all of the vital nutrients.*

How the World Eats

RIGHT *This grain bin is bursting with extra grain, while many people in the world are starving. Lack of food is not always the problem. For many countries, the problem is how to pay for the food.*

OPPOSITE *To grow a healthy crop like this yellow zucchini, good soil, sunshine, and plenty of water are needed.*

If you look around the world, you will find that people eat very different foods. People in rich countries such as the United States and Britain eat many different foods from all over the world. Many eat too much sugar and fat. They may also eat more protein than they need.

You have probably seen pictures on television of people dying of hunger when a **famine** is in the news. A famine happens when a country does not have enough rain to grow food, or cannot feed its people for some other reason, such as war. Thousands of people starve to death. We can see dreadful pictures on television, and we may give money to buy food for the people who are suffering.

But many people in poor countries are *always* hungry. They may not die of hunger, but they never have enough to eat. They suffer from diseases caused by too little food and not enough nutrients. Children who never eat enough protein get a disease called **kwashiorkor**. Their stomachs swell and they have little energy. They may die if they are not given protein. Any minor illness can kill them because their bodies are too weak to fight it off.

Where to Get Your Protein

If you eat meat, small helpings each day will give you enough protein. Remember that too much protein is not good for your body and too much meat may not be good for your planet.

Fish, like the white meats of chicken and turkey, is high in protein but has little fat. Most white meats are better for you than red meats.

Milk, cheese, yogurt, and other dairy products also come from animals. So do eggs. They are all rich in protein. You do not need to eat much of them to get enough for your daily needs.

On the whole, foods that come from animals give more complete proteins than foods that come from plants. Some groups of plant foods do not have one EAA, some groups are short of another. But if you put foods from different groups together, you can get all of the EAAs at once.

You can get all of your daily protein from plant foods, if you wish. If you don't eat meat, you will be eating a **vegetarian diet**. You can learn to make delicious meals that will give you all of the EAAs.

Plants that are full of protein include cereals and grains, legumes, seeds and nuts.

LEFT *Milk is rich in protein and vitamins. It also contains a lot of fat.*

BELOW *These colorful grains, beans, and peas are packed with protein.*

Some unusual foods to start with:

Whole wheat berries. These are delicious. Buy whole wheat, or other grains, at a health food store. Soak them in water for a few hours before cooking. To cook, use about 2 cups of water to each ½ cup of grain. Ask an adult to help you cook the berries.

To get good quality protein, try combining
legumes with cereals
seeds and nuts with cereals
seeds and nuts with legumes
Seeds and nuts include sunflower seeds and peanuts. Legumes include beans and peas. Cereals include soybeans and rice.

Some of the recipes that give you complete protein will not be new to you. A peanut butter sandwich combines cereal protein in the bread with protein in the peanuts. By eating these two proteins at the same time, you get a better quality protein.

ABOVE *How many of the beans, peas, and grains can you recognize?*

Bring the water to a boil in a large pan. Add the wheat and turn down the heat. Simmer, or cook gently, with a lid on for 1½ hours or as long as it takes to soften the grains. They won't get really soft like cooked rice. Add a little salt when cooked. Use them in place of noodles in a recipe.

Bulgur. This grain product is common in many parts of the world. It is partly prepared when you buy it. The wheat has been cooked in water, dried, and cracked before it was packed. It doesn't take long to cook. The easiest way to prepare it is to pour 1 cup of boiling water on ½ cup of bulgur in a large bowl. Simply leave it to swell and soften. Ask for help with the boiling water.

Legumes. You can buy most of the dried beans and peas already cooked and stored in cans. But you can also cook them yourself. Most legumes should be soaked in water overnight, or brought to a rapid boil and let stand for an hour, before cooking. This softens them so they cook more quickly.

Growing Your Own Sprouts

When seeds sprout, they are rich in many nutrients, especially protein. You can easily grow your own sprouts at home.

You will need:
An empty glass jar
A piece of muslin or cheesecloth
A rubber band
Some seeds (alfalfa seeds, chickpeas, mung beans, or soy beans)

1. Soak 2 to 3 tablespoons of seeds in warm water overnight.
2. Rinse the seeds in cold water and drain them.
3. Put the seeds in the jar and cover the top with muslin or cheesecloth. Fasten the cloth onto the jar with a rubber band.
4. Store in a warm, dark place. Rinse in cold water 2 or 3 times a day until the sprouts are long enough to eat. This takes 3 to 5 days.
5. Rinse, drain, and eat.

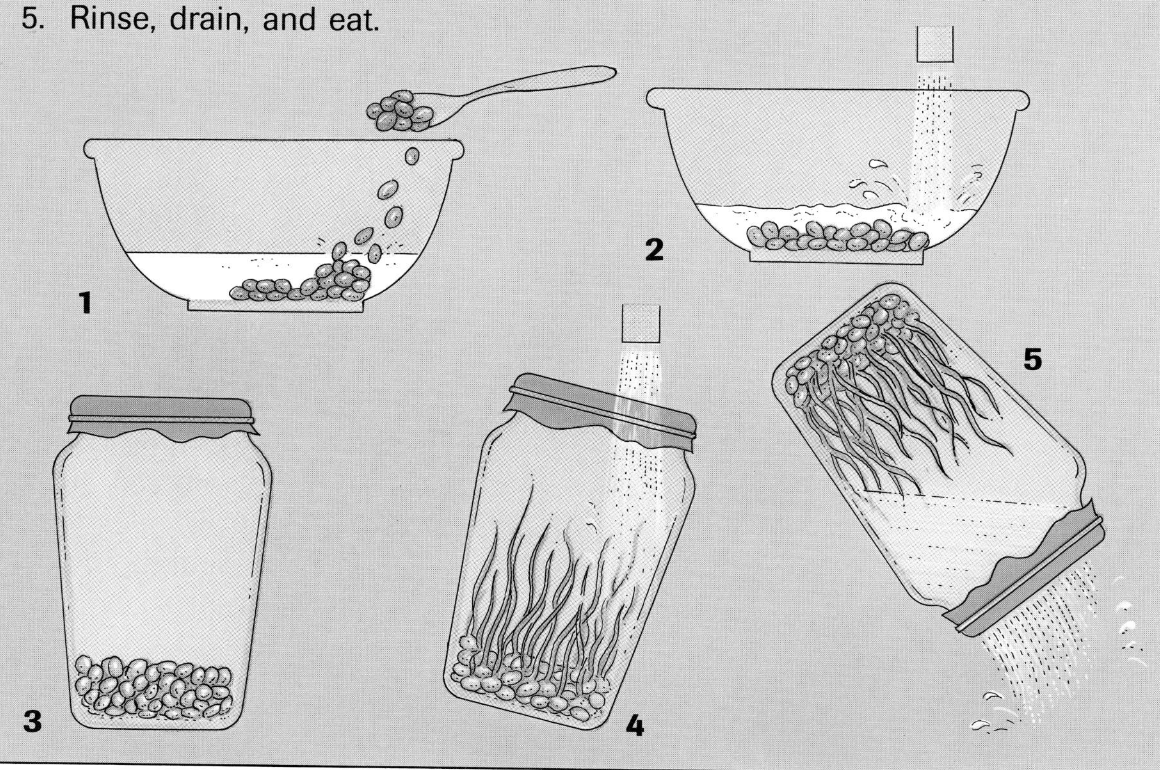

Your Daily Protein

Try to find out how much protein you're eating. You will need to keep a note of everything you eat in one day. Plan ahead. You will need a notebook, scales, and measuring cups to measure your food, and help from whoever cooks it. The packages or cans can be very helpful. They will list on the label how much protein there is in each serving. You can also look in the library for a book with more facts about protein in food. There are books that list the nutrients in foods sold in restaurants.

Whenever you can, measure how much of each food you eat that day. Guess the amounts for foods you cannot measure. Write down the amounts. Using books and the package labels, try to find out how much protein there is in what you have eaten. Write down the protein amounts. What is the total? How does it compare with the amount in the table on page 15? Perhaps you should eat more fruits and vegetables, and less food that is rich in protein.

Does most of your protein come from animals? How much comes from plants? Did any of your meals combine protein foods from the different groups on page 25?

To the right is a list of meals that combine proteins. Some of them you'll know. Others may sound strange at first. But they are all worth trying. You can add more of your own. Many cookbooks have good recipes using different mixtures of protein.

You have choices to make every day about what you eat. As you get older, you'll have even more choices. Your choices can help make ours a healthier earth. Cooking and eating healthy foods can help you live longer. Good luck with the food of the future!

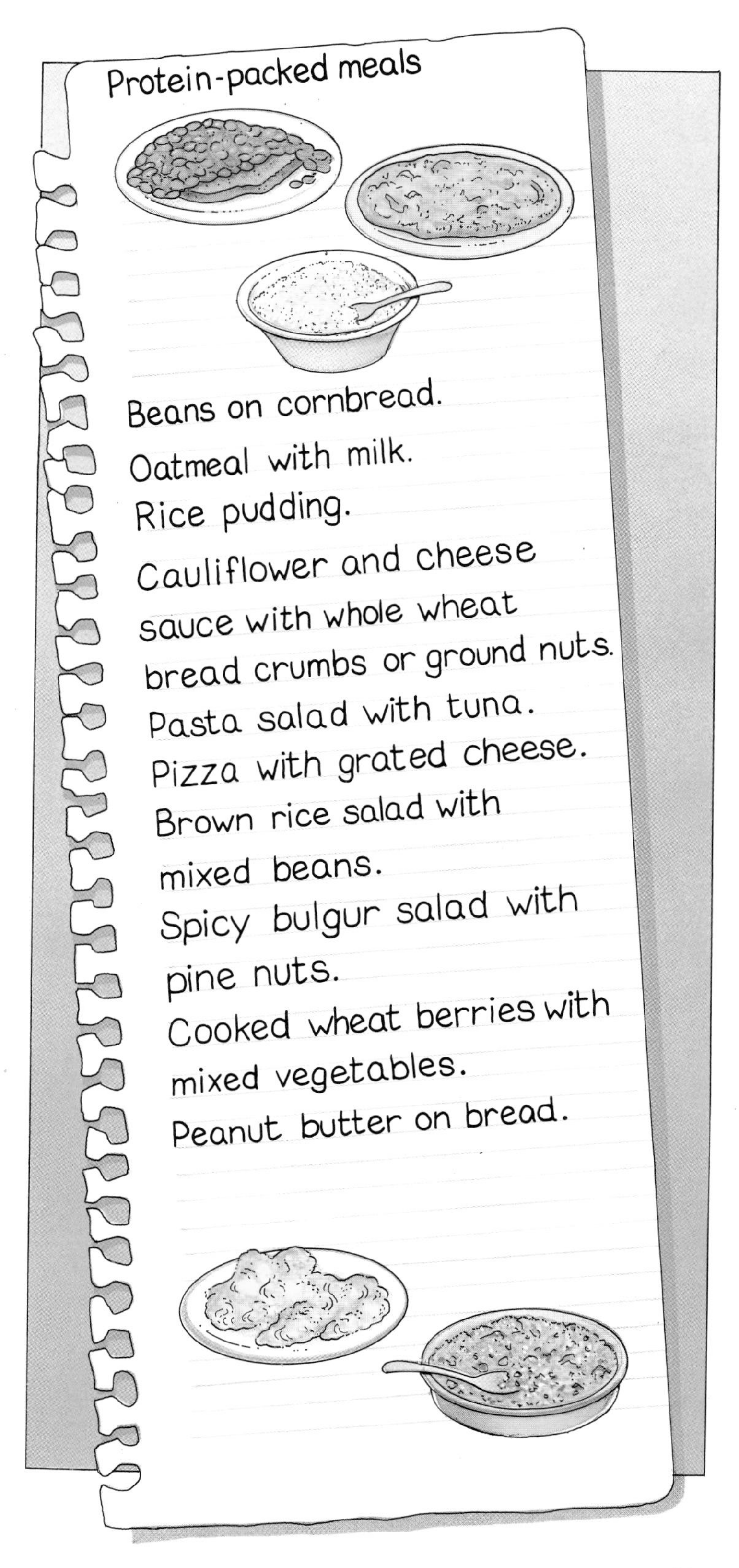

Glossary

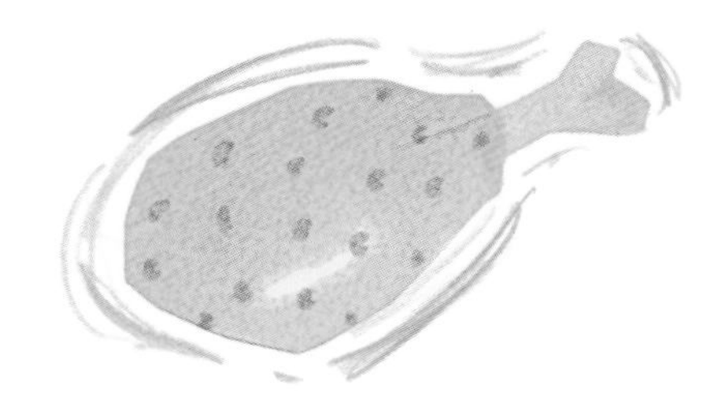

Amino acids Parts of a protein

Carbohydrates Nutrients such as sugars and starches that supply almost half of the energy we get from food

Carbon dioxide One of several gases in the air we breathe

Cell A tiny part of a living thing. It cannot be seen without a microscope.

Diet The things a person eats

Energy Power made in your body from the food you eat and the chemicals in your body

Essential amino acids The amino acids that our bodies cannot make and that we must get from foods we eat

Famine A shortage of food

Fat An important nutrient that supplies more energy than protein or carbohydrates. The body has other important uses for fat as well.

Kwashiorkor A disease of children caused by a diet that has too little protein

Minerals Substances, such as calcium, that our bodies need to get from food

Nitrogen A gas that makes up most of our air

Nutrients The things our bodies need from food to remain fit and healthy

Vegetarian A diet made up of vegetables, fruits, grains, and nuts, but no meats

Vitamins Substances, such as vitamin C, that our bodies need

Books to Read

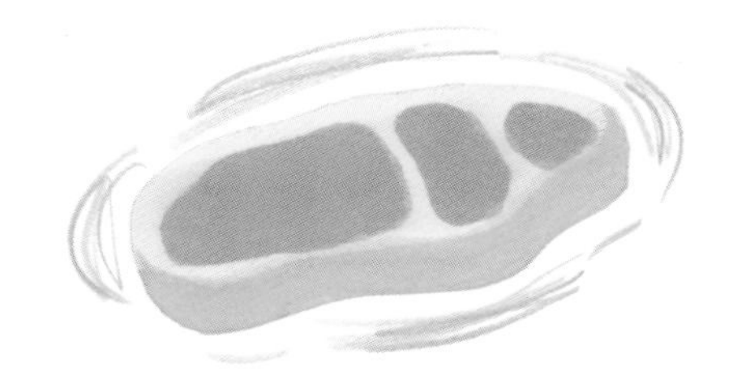

Beans and Peas by Susanna Miller (Carolrhoda Books, 1990)
Fish by Elizabeth Clark (Carolrhoda Books, 1990)
Food by David Marshall (Garrett Education Corporation, 1991)
Meat by Elizabeth Clark (Carolrhoda Books, 1990)

Metric Chart

To find measurements that are almost equal

WHEN YOU KNOW:	MULTIPLY BY:	TO FIND:
AREA		
acres	0.41	hectares
WEIGHT		
ounces (oz.)	28.0	grams (g)
pounds (lb.)	0.45	kilograms (kg)
LENGTH		
inches (in.)	2.5	centimeters (cm)
feet (ft.)	30.0	centimeters
VOLUME		
teaspoons (tsp.)	5.0	milliliters (ml)
tablespoons (Tbsp.)	15.0	milliliters
fluid ounces (oz.)	30.0	milliliters
cups (c.)	0.24	liters (l)
quarts (qt.)	0.95	liters
TEMPERATURE		
Fahrenheit (°F)	0.56 (after subtracting 32)	Celsius (°C)

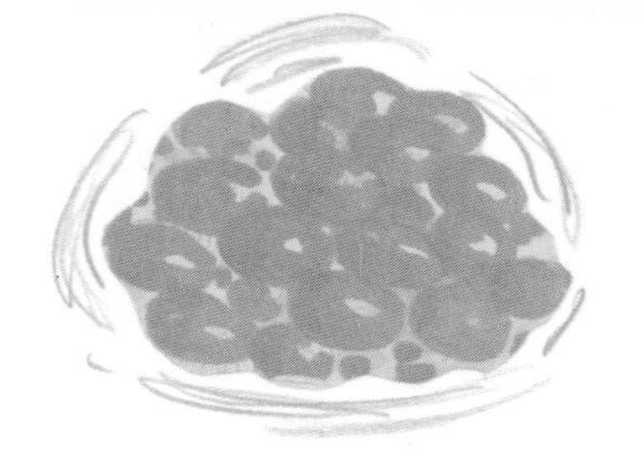

Index